Jean SHONGO YONGA
Nannethe KAWESI LOKOKISA

The issue of protected area management

Jean **SHONGO YONGA**
Nannethe **KAWESI LOKOKISA**

The issue of protected area management

THE CASE OF THE DEMOCRATIC REPUBLIC OF CONGO

ScienciaScripts

Imprint

Any brand names and product names mentioned in this book are subject to trademark, brand or patent protection and are trademarks or registered trademarks of their respective holders. The use of brand names, product names, common names, trade names, product descriptions etc. even without a particular marking in this work is in no way to be construed to mean that such names may be regarded as unrestricted in respect of trademark and brand protection legislation and could thus be used by anyone.

Cover image: www.ingimage.com

This book is a translation from the original published under ISBN 978-620-6-71572-6.

Publisher:
Sciencia Scripts
is a trademark of
Dodo Books Indian Ocean Ltd. and OmniScriptum S.R.L publishing group

120 High Road, East Finchley, London, N2 9ED, United Kingdom
Str. Armeneasca 28/1, office 1, Chisinau MD-2012, Republic of Moldova, Europe
Printed at: see last page
ISBN: 978-620-8-05402-1

0. Prelude

At the end of this in-depth planning study

of protected areas, we would like to express our sincere thanks to Professor **LOHAKA DJONGA**, Doctor of Science and leader of the seminar, for the didactic clarity of his presentation, his scientific rigor and his concern for training.

During his teaching, he made an enormous and profitable effort

using numerous examples to give concrete form to his most fundamental explanations. To develop in the listeners a spirit of analysis, comparison and initiation to the understanding of lesser. After providing a more thoughtful service, practical work was given to listeners with the real aim of cementing and concretizing the object of his teaching.

Our work is subdivided into three chapters, excluding the introduction and

Conclusion.

- The first chapter deals with the history and definitions of protected areas.

- The second chapter deals with the management of protected areas,

- The third chapter deals with the planning process (identification of stakeholders, prioritization of planning actions and sustainable financing).

CHAPTER I: HISTORY AND DEFINITIONS OF PROTECTED AREAS

1.1 History and overview

Nature protection is very old, dating back more than 2,000 years, and is relatively universal: the creation of the very first protected area is generally attributed to the emperor Ashoka around 252 BC in India, who built a sanctuary around his palace to preserve flora and fauna.[6]. In Europe, hunting areas have been reserved for the elite for centuries, while Pacific societies have placed "taboos" on certain areas, and "sacred forests" have been, and sometimes still are, subject to religious prohibitions in Africa.

In the modern era, one of the first non-profit protected areas was the forest of Fontainebleau: in the **19th** century, this hunting reserve and later royal marine reserve (never exploited) was the last place in the Paris region where very old trees could be admired.[e] century, this was the last place in the Paris region where very old trees could be admired, much to the delight of walkers and, in particular, artists such as the painters of the Barbizon School.

An imperial decree of April 13, 1861 created the Fontainebleau "artistic reserve", which was increased to 1,094 ha and finally to 1,693 ha from 1892 to 1904, making it the very first officially instituted nature reserve in history.

In the modern era, one of the first non-profit protected areas was the forest of Fontainebleau: in the **19th** century, this hunting reserve and later royal marine reserve (never exploited) was the last place in the Paris region where very old trees could be admired.[e] century, this was the last place in the Paris region where very old trees could be admired, delighting walkers and, in particular, artists such as the painters of the Barbizon School. When the forest began to be exploited, these artists, led by Théodore Rousseau, founded the Société des amis de la forêt de Fontainebleau to protect the forest.[8].

An imperial decree of April 13, 1861 created the Fontainebleau "artistic reserve", which was increased to 1,094 ha and finally to

1,693 ha from 1892 to 1904.[10]making it the very first officially established nature reserve in history.[11].

The movement took another step forward in 1864, when the U.S. Congress awarded the State of California the management of a small area for public use. This valley became Yosemite National Park in 1890, under the impetus of wildlife protection pioneers John Muir and Robert Underwood Johnson. The first site to bear the title of national park was Yellowstone, created in 1872 in the United States. At the same time, the first steps were taken to create Blue Mountains National Park and Royal National Park in what was then the British colony of Australia. Similar initiatives emerged in Canada, New Zealand and South Africa.[7]

1.2 Terminology

In French, the term espace protégé or espace naturel protégé[3] or "espace naturel protégé" sometimes replaces the term "aire protégée".

A Marine Protected Area (MPA) is defined as a geographical area whose protected status includes all or most of a marine zone.[4].

Growth in the area classified as MPAs is only 5% per year, so at this rate we'd have to wait fifteen years to hope for a doubling in the area theoretically protected.

A transboundary protected area is one where the protected geographical area aggregates and ensures a continuum of protected areas on either side of the borders of several countries (usually this area exists formally through an international agreement that defines coordinated management rules between the aggregated national protected areas).

1.3 Interests of protected areas

In China, between 1975 and 1985, five "National Parks" were created specifically for the protection of giant pandas, including the Wolong Nature Reserve.[17].

Protected areas have multiple functions, and their promoters may be motivated by the need to protect pristine nature for its intrinsic value.

Since the 1970s-1990s, various authors (particularly in the West) have attributed to them the capacity to provide numerous ecosystem services or amenities: natural water purification, in the case of wetlands, or atmospheric carbon sequestration in the case of forests, as part of the fight against global warming.

More directly for the economy, these natural areas can serve as leisure zones and attract significant touristactivity, while other areas are dedicated to sustainable agriculture.

Finally, the existence of certain protected areas is justified by the history of a place or the cultural practices associated with it.

The first protected areas were often classified for their beauty, as shown by the examples of Yellowstone in the United States and the Fontainebleau forest in France.

An analysis of the sites listed under the Ramsay Convention shows that while in 1993 the majority were listed on the basis of the presence of remarkable species of fauna (in this case, waterbirds), the emphasis has increasingly shifted to human development criteria.[18]

The author of the study argues that "conservation" is not the priority of developing countries, and that this shift in the convention's priorities may have been intended to encourage more new countries to ratify it.

In practice, according to the research carried out by G.Baldi and his team in 2016, the choice of where to locate protected areas has largely been based on the opportunities offered by their remoteness from population centers and low population density. This is particularly true in North America, Australia and New Zealand, and in Latin America and the Caribbean.[19]

1.4. Scientific, economic and ethical justifications (reasons).

Note that :

On the basis of the foregoing, we successively identify the scientific, economic and ethical justifications (reasons) for conservation, after noting the role or importance of protected areas (PAs) in **biodiversity conservation.**

A. Scientific justifications (reasons).

For example;

-In some parts of Paris, for example, you could **admire** very old trees, which delighted walkers and artists like those at the school.

-Westerners in particular attribute to protected areas the capacity to provide numerous ecosystem services or amenities, such as natural water purification, wetlands or atmospheric carbon sequestration for forests, as part of the fight against global warming.

In the case of nature reserves (cat_1) on the geomorphology entity;

In the case of management reserves (cat_4) on habitats and species;

- In the case of natural resources, the aim is to conserve the various types of natural ecosystem in order to preserve processes.

B. Economic justifications (reasons) For

example :

- To preserve flora and fauna;
- To serve as a leisure area and attract significant tourist activity;
- For other areas dedicated to sustainable agriculture ;
- The existence of certain PAs is justified by the history of a place (or the cultural practices linked to it);
- For the Ecological (cat_2) and/or Aesthetic ;
- For natural monuments (cat_3) protection is justified by their uniqueness or rarity;

- For landscapes and seascapes (cat$_5$) on aesthetics for leisure and tourism purposes.

C. Ethical reasons

For example: taboos in Pacific societies because in some places "sacred forests" were or still are sometimes the object of religious prohibitions in Africa.

1.1. PROTECTED AREAS (AP)

1.1.1. Definition of a protected area (PA)

A protected area is a delimited geographical space, managed and intended for the long-term conservation of the natural resources it contains, as well as the ecosystem services or amenities (e.g.: natural water purification) it provides.

Almost every country in the world has established protected areas. Several international programs seek to protect nature and encourage states to classify natural areas. Although natural resources have always been managed by human societies, the first protected areas were created in the United States at the end of the XIX[e] century, before concern for the environment became global at the turn of the 1970s, then in Rio in 1992 with the Convention on Biological Diversity (CBD), and the movement spread across the planet.

In 2018, a scientific consensus establishes that protecting habitats and species is the best way to defend biodiversity, provided that protection is real in parks and reserves. According to Jones *et al.* (2018), the most recent and comprehensive global mapping of human pressures shows that while there has been an increase from 9% to 15% of protected emerging land between 1992 and 2018, six million square kilometers (or 32.8% of protected land) are nevertheless subject to intense human pressure[2]. More than 55% of protected areas prior to CBD ratification (1992) have since been subject to increased human pressure; only large, strictly protected areas are potentially effective in some countries, mainly in the southern hemisphere.[2]

1.1.2. Criteria for determining a PA.

The ecological criteria used by experts to determine a protected area are endemism rates, diversity and the presence of a large number of endangered species.

- **THE CASE OF PROTECTED AREAS IN THE DEMOCRATIC REPUBLIC OF CONGO**

1.1.3. Protected area (PA) categories

IUCN has defined **six categories** of main types of protected areas.

The example of the DRC's protected areas can be cited.

- **Category 1:** integral nature reserves (nature reserves).

These are ecosystems of national or international scientific importance. They contain fragile habitats and species, sometimes even entire communities, that are under threat. Artificial disturbance, tourism and public access are generally prohibited in order to ensure that fundamental ecological processes are not disrupted.

-Category 2: National parks (relatively large territories)

A park is made up of one or more types of contiguous ecosystems, which may or may not have been transformed by human activity, and whose living communities, habitats and geomorphological sites are of scientific, educational, recreational or aesthetic interest, and where the public authorities have taken all the necessary measures to prevent or eliminate as quickly as possible any exploitation or occupation over its entire surface area, in order to ensure effective respect for the ecological, geomorphological and/or aesthetic features that justified its creation.

Consequently, the exploitation of natural resources must be prohibited in a national park. The only exceptions allowed in national parks concern sites intended to accommodate tourists, limited in number and surface area, and the minimum roads required for access. A zoning system to avoid potential conflicts of interest between tourist access and conservation.

Category 3: Natural monuments

These are areas containing one or more special features of exceptional national importance, whose protection is justified by their uniqueness or rarity.

They may be of cultural importance, especially historically.

Natural monuments do not occupy large areas, but rather relatively small ones. Although they are open to the public, natural monuments must be preserved from artificial disturbance.

-Category 4: Managed nature reserves (also known as habitat and species management reserves)

They respond to the need for targeted protection, and therefore imply active intervention in contrast to the former. They correspond to the former "special" or "specialized" reserves (botanical reserves, hunting reserves, sanctuaries, wildlife sanctuaries, etc.), since as a general rule they represent in one way or another an exploitation of their natural biological resources.

-        **Category 5:** Protected landscapes and seascapes

There are 2 main types of protected areas in this category.

-        Landscapes with special aesthetic qualities resulting from the interaction of man and nature;

-        Natural areas that have been intensively developed by man for leisure or tourism purposes.

The disadvantage of these protected areas is that they offer little protection for the habitats and biodiversity they contain, and only prohibit industrial construction and urbanization, which would tarnish the landscape and disfigure the traditional habitat and architectural heritage.

-        **Category 6:** Managed natural resource areas

These are very large, relatively isolated, uninhabited or sparsely populated areas that are difficult to access, but which may be under pressure to colonize (often the case in the third world).

These are areas that have generally been little studied or assessed, and where the consequences of agricultural, forestry or mining development are poorly understood.

These areas require rational management of their resources. The pressure exerted by local or migrant populations to exploit them must be controlled.

Two main categories of protected areas created by **UNESCO**.

***Biosphere reserves**

This category of protected area, which is particularly important for conservation, was created as part of **UNESCO**'s MA (Man and Biosphere) program **in 1970** to conserve at least one representative area of each major ecosystem type in the world and to create a network of biosphere reserves.

In reality, these biosphere reserves often overlap partially or entirely with other existing protected areas that fall into the six basic categories defined by the IUCN (1 to 5).

The essential objectives of management are therefore to conserve the various types of natural ecosystems in order to preserve fundamental ecological processes and the genetic diversity of the plant and animal communities that inhabit them.

An essential purpose of biosphere reserves is to serve as a milestone, a witness to the evolution of ecosystems over time, whether spontaneous or man-made.

Lastly, their purpose is to enable fundamental or applied research in the field of Ecological Sciences.

They are areas for environmental monitoring. One of the last essential characteristics of the biosphere is that their status provides for a zoning system to guide their management. This zoning comprises 3 generally concentric zones:

- A central area, intended as a virtually complete reserve and as natural as possible, a buffer or strictly delimited zone and a peripheral transition area, a stable cultivation area in which the farming techniques used should be as ecologically undisturbed as possible, and in which the cooperation of local populations is systematically sought.
- Natural **world** heritage sites

On the proposal of a State, a signatory party after endorsement by the International World Heritage Committee, areas of outstanding universal value may be designated as World Heritage sites. These are often already protected areas falling into categories 1 to 5, and are also created to ensure ongoing environmental research and monitoring. Although the primary reason for the reaction to the creation of World Heritage sites is the exceptional character for human civilization of the ancient remains or monuments they contain, the erection of the latter in this category of protected area may in part be linked to their characteristic as natural monuments.

CHAPTER II: MANAGEMENT OF PROTECTED AREAS

A Protected Area (PA) is defined as a delimited terrestrial, marine, coastal or aquatic territory whose components are of particular value, notably biological, natural, aesthetic, morphological, historical, archaeological, cultural or spiritual, and which requires, in the general interest, multifaceted preservation. It is managed with a view to protecting and maintaining biological diversity, conserving the particular values of natural and cultural heritage and ensuring the sustainable use of natural resources, thereby contributing to poverty reduction;

Protected areas are just one of the sectors and professions that have seen an increasing demand for collaboration with a wide range of stakeholders in recent years.

Environmental and natural resource management has moved from a top-down regulatory style to one of close and diverse partnership and collaboration between management agencies and local communities, resource users, other management agencies, non-governmental organizations (NGOs) and the private sector. This is in line with broader arguments about the role of citizens, power-sharing and participation in political and strategic decision-making, as well as the shift from government direction to more inclusive, multi-stakeholder governance.

2.1. Purpose of a PA development plan

Planning is the process by which stakeholders (anyone who may be affected by an area's environment, or who may have an interest in its development, even if indirectly) come together to consider and determine how to manage the resources in a given place for the benefit of present and future generations.

2.2. Components of a PA development plan

2.2.1. PA environment to be managed

This part concerns everything that surrounds the PA and can have an impact on its development.

2.2.1.1. Situation or legal status in relation to the region or country

- Is there a national resource protection plan of which the PA should be part? A network of PAs to which it should belong?
- Is the national plan based on a global conservation strategy proposed by IUCN?

2.2.1.2. Use of adjoining land

- For **operation**

  - For farming purposes

  - For forestry or hydroelectric development to generate electricity

  - For the construction of infrastructure serving

  -industrial and commercial development.

  - urbanization (markets, schools, hospitals, etc.)

  - outdoor recreation and tourism

  - For the construction of transport system infrastructures providing access to the PA

  - Roads (vehicles)

  - Tracks (trains)

  - Runways (aircraft)

2.3 Characteristics of a protected area

2.3.1. Demographics

The aim is to identify trends in order to anticipate changing loads on equipment, services and investments.

2.3.2. Cultural features

- These are :

✓ Important historical sites (remains of prehistoric civilizations, scientific sites): places, land, buildings and objects of historical value.

✓ Predominant group with distinctive social and religious customs (rites, ceremonies).

2.3.2. Physical characteristics

A characteristic is a quality that identifies someone or something, and physical is a term that can have several meanings.

In this case, the adjective indicates what belongs to or relates to the physical constitution.

As far as we are concerned, we will base ourselves on :

✓ Demarcate and describe boundaries, using natural features wherever possible,

✓ Identify topography, waterways and specific physical features

✓ Compiling maps / satellite images

2.3.3. Ecological features

It's all about easy-to-understand features related to packaging eco-responsibility;

For us, these are the combinations of processes and benefits, systemic Eco-services that characterize the area at any given time. Our approach is based on identifying the characteristics of the protected area.

Identifying the characteristics of the protected area remains a relative matter:

- The main wildlife resources?
- Wildlife migration and movement corridors
- Rare and under-represented plant formations

- Other floral and faunal resources of great importance to the protected area,
- Describe essential ecological processes within the protected area and interactions with areas outside the protected area.

2.3.4. Socio-economic characteristics

It is the overall claim to be able to describe and explain economics and economic action in a more realistic way to :

- Identify villages, cultural and spiritual resources, trails, transportation routes, major economic centers in and around the landscape, agricultural activities, hunting and fishing areas, and subsistence timber extraction areas.

2.3.5. Stakeholder characteristics

✓ Identify all PA stakeholders (including populations living outside the PA or who have been displaced)

✓ Convene stakeholders to define PA objectives. It may be necessary to meet several times to decide on objectives.

✓ Outline the objectives of the PA and, if possible, list them in order of priority.

✓ Describe the possibilities for achieving each objective and any difficulties.

✓ Identify which resources and areas of the PA are used for subsistence purposes and which are used for commercial purposes (species hunted or gathered and intensity).

✓ Draw up a map of wood species or mineral deposits that are desirable for economic reasons and that could be exploited in the future.

✓ Identify current legal and illegal uses of PA resources

✓ Describe other existing economic activities that rely on the PA, such as tourism.

✓ Facilities Identify existing infrastructure: roads, administrative buildings, airstrips, tourist lodges, etc.

✓ Describe the impact of surrounding land uses

✓ Describe known threats to the above resources and known trends affecting them

✓ Consider possible future challenges and new or changing influences on PA

✓ Assess the presence of the national planning authority in the PA and its ability to implement the plan and enforce laws.

2.3.6. Characteristics of stakeholders

✓ Characterize the PA, specifying its features and known attributes.

✓ Keep descriptions objective and brief.

✓ Use tables and maps wherever possible to list PA natural resources and describe physical, ecological and socio-economic conditions.

6°) Special features

This refers to any phenomenon affecting the entire protected area, or one or more parts of it, which could have an impact on its use, mission or development (landslides or subsidence, mudslides, etc.).

2.4. Unique value of the protected area

✓ A PA plan is an organized management tool for the PA, and serves as a guide to whether the status of the PA (at given points on a timeline) is still consistent with its purpose at local, national, regional (including CBFP) and global levels.

✓ In some cases, the plan also provides a strategy for obtaining official recognition of the protected area and recording its original purpose. As such, a presentation of the protected area and the resources it contains, explaining its unique value, is a useful way to begin the plan.

✓ Ensure that the description is brief and highlights the essential features of the PA, as well as its role in the overall landscape context, that have helped to explain why it has been classified as a PA or why it should be.

✓ A protected area could be an essential link in the network within the landscape, for example; the roles of the protected area in the overall landscape should be clearly explained in the landscape plan.

✓ The management plan is not suitable for an in-depth or complicated discussion of the resource.

Protected area activities are planned on surrounding land and will provide a source of employment for local communities. The following parts of the plan will explain the above points in more detail, so highlight in this section the important elements - the key points we want to make known about the protected area.

Our tasks will be to:

✓ Identify and describe the unique value of this protected area.

✓ This serves as an introduction to the management plan, and should be kept brief and concise.

This section should clearly and briefly answer the question "Why is this land a protected area?"

2.5. Desired conditions

The desired conditions for a protected area should indicate what the area should look like, and the benefits it will provide indefinitely into the future.

Describe the desired conditions for the protected area, relating them to the desired conditions and objectives for CBFP Landscapes, as well as to national objectives for protected areas.

The desired conditions will need to take into account the unique qualities of the protected area and indicate how it can contribute to meeting the conservation purpose for which it was created, the needs of stakeholders and the goals of the CBFP, namely to establish sustainable natural resource management practices throughout Central Africa, thereby promoting sustainable economic development and alleviating poverty for the benefit of the people of the region and the global community (State of the Forest Report 2005, p.2).

The desired conditions for the protected area will provide context and guidance for the rest of the planning process. Planning and management of protected areas in Central Africa: (US Forest Service Guide)

Version 2.0 Page 17)

Identify the context, role and purpose of this PA within the network of PAs and other land-use categories (macro-zones) in the CBFP landscape, in the country as a whole, at Congo Basin level and globally. If the country in which the protected area is located has adopted a national vision or set of goals for its protected area network, then the desired conditions should take these into account and reflect this national vision in the development of the country's protected areas.

Some questions to consider in developing the desired conditions include:

- What's unique about the protected area and why is it famous?
- How does this protected area differ from the surrounding land?
- What do planners and stakeholders want the protected area to look like ecologically?
- How should the protected area contribute to the social well-being of the region and its inhabitants?
- What resources need to be preserved or protected?
- What category of protected area does the plan seek to establish?

- To what extent will the protected area contribute to biodiversity conservation, heritage, local communities or economic development and poverty alleviation?
- How will the PA meet the goals of CBFP Landscapes, and to what extent does it serve as an indispensable link in the PA network?
- What should it look like, and what should it deliver for an indefinite future period?

The desired conditions will have to be worked out by consensus with the stakeholders.

Consequently, this part of the plan will also have to indicate who was involved in drawing them up.

Seek to develop desired conditions that take into account economic and social considerations, as well as the characteristic roles of the PA and its contributions to ecosystems.

Specify the desired conditions for the PA at the beginning of the document to provide context and direction for the rest of the plan. Most projects and activities are developed specifically to achieve or preserve one or more of the desired conditions and objectives of the plan. It should not be expected that every project or activity will contribute to all the desired conditions or objectives in every case, but Desired Conditions Desired conditions set the general direction for the protected area over a long period of time.

Desired conditions set ideal goals for what the protected area should be, what it should protect and who should benefit from it.

Some examples of desired conditions are shown below:

1) Forest elephant populations are re-established to occupy areas of historic presence, at historic densities, to ensure ecological processes, such as the dispersal of native tree seeds and the creation and maintenance of forest basins used by a host of other species.

2) Bonobo strongholds are protected from exploitation and ensure the repopulation of the species from source populations.

3) Ecotourism is established as a small, sustainable source of income for local communities and contributes to the long-term conservation of forest ecosystems. Planning and Management of Central African Protected Areas (US Forest Service Guide Version 2.0 Page 18) only to a selected subset. PA management plans will need to formulate the desired conditions, the activities proposed to achieve these conditions and whether these conditions and objectives are moving in the right direction. The desired conditions may only be achievable in the long term.

If it becomes clear that the desired conditions cannot be achieved, or that they no longer apply to the long-term multi-purpose management foreseen in the plan, update or revise the plan:

1) Convene the PA planning team and stakeholders to develop the desired conditions for the PA.

2) Define desired conditions that are widely accepted and that seek to preserve the PA's unique characteristics and significance, improve the conditions of the PA's resources and promote livelihood opportunities for those who depend on the PA's resources or for those who could benefit from them. The desired conditions should reflect the visions or goals established by the government for all PAs in the country in question.

3) Indicate those who participated in the development of the desired conditions, so as to clarify the desired conditions they represent.

2.6. Objectives

The management objectives set out the main principles that are essential to the effective management of the PA.

Objectives are particularly important because they support the desired conditions and, more specifically, they describe the expected result for a given element, attribute or condition of the PA. (Example: within 10 years, this PA will support and preserve diverse and sustainable populations of local fauna, fish and plants).

Other, more specific objectives may be set for species or ecosystems that give cause for concern.

- The objectives should be sufficiently numerous to ensure that as many PA issues as possible are adequately addressed.

- Targets don't have to specifically state how they will be achieved, but they do have to be achievable.

- Objectives must be unequivocal, measurable and accompanied by a timetable.

- It is essential to involve stakeholders in the development of objectives, since different stakeholders sometimes disagree on which activities are compatible or not with the desired conditions.

- It won't be possible to satisfy all stakeholders, but planners will need to properly assess the objectives of different stakeholders and find answers to their controversial or contradictory views.

- If necessary, the planning team may wish to apply conflict resolution methods, including negotiation techniques, to help resolve key disputes between stakeholders.

The term "issue" here refers to any topic concerning uses, threats, opportunities, activities, conflicts, etc., related to the protected area.

Where possible, list objectives in order of priority. Objectives for the protected area plan may be based on the following themes, but will be specific to the site in question: Conservation of habitat and species (fauna and flora) Promotion of scientific research Preservation of social and cultural features Education and training Community participation and development Income generation

Development of ecotourism Ecological services

For each goal, explain the opportunities and challenges involved in achieving it. For example, poverty and a precarious economy may continue to result in hunting pressure on endemic species of concern. Wherever possible, incorporate the wishes of communities and stakeholders, or explain how these needs are taken into account on surrounding lands within the landscape. We distinguish between two kinds of objectives: hypothetical and real.

***Examples of hypothetical objectives* :**

- Clearly establish checkpoints and post guards on all roads approaching within 8 km of the protected area.

- Publish protocols to identify potential poachers.

- Patrol the protected area at regular intervals to detect poaching camps. Use the information gathered to develop strategies to combat poaching.

- Establish two wildlife observation sites within the protected area for tourists. Ensure that new human settlements within 8 km of the protected area have sufficient food sources so that they do not rely on game living in the protected area for their livelihood.

Example of real objectives:

- Protect endangered and endemic species.
- Preserve flora and fauna protected by national laws.

- Promote the sustainable management of natural resources by traditional communities in the protected area.

- Establish a tourism site as a source of funding for sustainable development.

- Promote conservation education and awareness-raising among local populations. (Source: Development plan and business plan for the Tanya nature reserve, Lubero territory, Batangi and

Bamate chiefdoms, North Kivu province, Democratic Republic of the Congo, 2008)

- Reconciling the interests of wildlife conservation with those of local populations, while respecting traditional customs and forestry and mining development.

- Prevent hunting incursions into forest concessions and reduce hunting pressure.

- Better manage park boundaries and establish a control system. Slow down or prevent cross-border hunting for commercial purposes.

- Identify the context, role and purpose of this protected area within the network of protected areas and other land-use categories (macro-zones) in the CBFP landscape, in the country as a whole, at Congo Basin level and globally. If the country in which the protected area is located has adopted a national vision or set of goals for its PA network, then the desired conditions should take these into account and reflect this national vision in the development of the country's protected areas.

2.7. Guidelines

Guidelines can be conceived as a series of standards or rules that apply to the entire protected area, describing the activities that are permitted or prohibited.

Guidelines ensure that certain aspects of the protected area maintain their integrity and that various activities take place, or are prohibited, in such a way as not to compromise the valuable attributes of the protected area. In this section, name and describe the protected area guidelines that apply universally to the entire area.

Further guidelines will subsequently be drawn up for each micro-area; these guidelines will apply only within the micro-area for which they are intended. The guidelines will prohibit or permit specific activities or actions. These guidelines must also recognize customary use and access rights, and ensure resource use as recognized in other official texts (e.g. concessions, parks, etc.).

Protected area guidelines may address the following points; however, remember that simplicity is preferable and adopt only those protected area guidelines that are necessary to preserve its character and achieve the set objectives and desired conditions: Hunting and fishing: specify whether these activities are permitted and, if so, specify for which species, on what dates, where, by what means and in what quantity (limit per person, per season or per day) and who may carry out these activities (local communities, recreational hunting or recreational fishing).

Logging: specify whether logging is permitted and, if so, a number of guidelines to govern operations that are ecologically sustainable.

Here too, the guidelines must specify who can harvest the wood, in what quantities, when, where and which species (the guidelines can indicate which species can be harvested, or which cannot, whichever is simpler).

- Collection of non-timber products: specify whether this is permitted and, if so, which species or products may be collected, where, when, in what quantities and by what method.

- Motorized vehicles: specify where they are allowed, when and what regulations apply (e.g. staying on roads reserved for vehicles).

- Non-motorized recreation: set the rules that apply to the various types of recreation that may take place in the protected area.

- Roads: include in the guidelines a map of existing, planned and closed roads. Decisions will have to be made as to which roads will be maintained and which will be permanently closed. There may be places where traffic will be discouraged to protect an aspect of the protected area, and places where road infrastructure will need to be improved to facilitate management or tourism.

- Determine the size of authorized vehicles and whether traffic volume will be controlled.

- Economic corridors: identify, map and define the acceptable use of the main corridors for the transfer of goods and services.

- Corridors can include roads, tracks, trails, waterways or other means of transporting goods and people for trade or economic transactions.

- Trails: as for roads, indicate which means of transport are permitted on each trail (e.g. pedestrian, bicycle, horse/mule, motorcycle).

- Infrastructure development: include a map of existing facilities, such as transmission towers, water pipes, power lines, buildings, etc., in the guidelines for achieving the desired level of infrastructure in the protected area.

Infrastructure guidelines will need to consider development in "built corridors", rather than in a haphazard way that risks increasing negative impacts, such as habitat fragmentation or detracting from the aesthetics of the protected area.

Fire: will fires be allowed and, if so, who can light them and under what circumstances?

- What are the rules for extinguishing fires?

- Cultural heritage resources: if there are any in the protected area, who will have access to them, when and what type of rituals, if any, will be practiced.

- Minerals and geology: will prospecting and extraction be authorized?

Tourist activities: who can bring tourists into the protected area, what permits are required, are guides required, what fees will be charged, is camping allowed, are night visits permitted, etc.?

Scientific research: what permits are required, what limits of environmental manipulation will be allowed. **Community rights and development**:

- Are there any villages that already exist within the protected area? Will they be allowed to stay, and if so, what resource use rights will they retain? How will revenues be shared with local communities?

- Will local residents receive preferential treatment for jobs related to the protected area?

If exceptions are provided for in a given guideline, the guideline should explicitly describe the circumstances under which the exemption will be granted, and specify the authority that will grant it. For example, if the capture or hunting of animals in the protected area is going to be prohibited by the guidelines, the author may consider drafting an exception authorizing the park administration to control certain species for management purposes, or to allow capture or killing for scientific research purposes, provided the corresponding permits are held.

It is also important to note that laws already in force in the country where the protected area is located may address certain issues or activities, and that the protected area will remain under the jurisdiction of these laws. In such cases, these laws will need to be referenced in the guideline; however, the guidelines may set out more stringent rules that complement the laws already in force.

Our tasks will be to :

1)   Name and describe the protected area guidelines, applicable throughout the protected area.

2)   Remember that simple guidelines are best. Each one should serve to achieve the objectives and desired conditions determined in advance. Describe any exemptions from the guidelines, specifying who can benefit and under what circumstances.

Chapter III: PLANNING PROCESS (IDENTIFICATION OF STAKEHOLDERS, PRIORITIZATION OF PLANNING ACTIONS AND SUSTAINABLE FINANCING).

3.1. Guidelines

Guidelines can be conceived as a set of standards or rules that apply to the entire PA, describing the activities that are permitted or prohibited within it. Guidelines help to ensure that certain aspects of the PA maintain their integrity, and that various activities take place, or are prohibited, in ways that do not compromise the valuable attributes of the PA. In this section, name and describe the PA guidelines that apply universally to the entire area. Additional guidelines will subsequently be drafted for each micro-area; these guidelines will apply only within the micro-area for which they are intended. The guidelines will prohibit or permit specific activities or actions. These guidelines must also recognize customary use and access rights, and ensure resource use as recognized in other official texts (e.g. concessions, parks, etc.).

PA guidelines can address the following points; however, remember that simplicity is preferable and adopt only those PA-level guidelines that are necessary to preserve its character and achieve the set objectives and desired conditions: Hunting and fishing: specify whether these activities are permitted and, if so, specify for which species, on what dates, where, by what means and in what quantity (limit per person, per season or per day) and who may engage in these activities (local communities, recreational hunting or recreational fishing). Logging: specify whether logging is permitted, and if so, a number of guidelines should govern operations that are ecologically sustainable. Here too, the guidelines should specify who may harvest timber, how much, when, where and what species (the guidelines may indicate which species may be harvested, or which may not, whichever is simpler). Collection of non-timber products: specify whether this is permitted and, if so, which species or products may be collected, where, when, in what quantities and by what method. Motorized vehicles: specify where they are allowed, when and what regulations apply (e.g. staying on roads reserved for vehicles). Non-motorized recreation: set the rules that apply to the various types of recreation, if any, that might take place in the PA. Roads: include in the guidelines a map of existing, planned and closed roads. Decisions will have to be made as to which roads will be maintained and which will be permanently closed. There may be places where traffic will be discouraged to protect an aspect of the PA, and places where road infrastructure will need to be improved to facilitate management or for tourism. Determine the size of vehicles allowed and whether traffic volume will be controlled.   Economic corridors: identify, map and define the acceptable use of key corridors for the transfer of goods and services. Corridors may include roads, tracks, trails, waterways or other means of transporting goods and people for trade or economic transactions. Trails: as for roads, indicate which means of transport are permitted on each trail (e.g. pedestrian, bicycle, horse/mule, motorcycle).

Infrastructure development: include a map of existing facilities, such as transmission towers, water pipes, power lines, buildings, etc., in the guidelines for achieving the desired level of infrastructure in the PA. Infrastructure guidelines will need to consider development in "built corridors", rather than doing so in an unmethodical way that risks increasing negative impacts, such as habitat fragmentation or detracting from the aesthetics of the PA.   Fire: will fires be allowed, and if so, who can light them and under what circumstances? What are the rules for putting out fires?

Cultural heritage resources: if there are any in the PA, who will be able to access them, when and what type of rituals, if any, can be practiced. Minerals and geology: will prospecting and extraction be permitted?   Tourist activities: who can bring tourists to the PA, what permits are required, are guides required, what fees will be charged, is camping permitted, are night visits allowed, etc.? Scientific research: what permits are required, what limits of environmental manipulation will be allowed.

Community rights and development: are there any villages that already exist in the PA, will they be allowed to stay and, if so, what resource use rights will they retain? How will revenues be shared with local communities? Will local residents receive preferential treatment for PA-related jobs?

If exceptions are provided for in a given guideline, the guideline should explicitly describe the circumstances under which the exemption will be granted, and specify the authority that will grant it. For example, if the capture or hunting of animals in the PA is going to be prohibited by the guidelines, the author may consider drafting an exception authorizing the park administration to control certain species for management purposes, or to allow capture or killing for scientific research purposes, provided the corresponding permits are held. It is also important to note that laws already in force in the country where the PA is located may address certain issues or activities, and that the PA will remain under the jurisdiction of these laws. In such cases, these laws will need to be referenced in the guideline; however, the guidelines may set out more stringent rules that complement the laws already in force.

A management plan is a document that sets out the management approach and objectives, and includes a framework for decision-making, to be applied in the protected area over a given period. It is now critical for protected area planning to consult as many stakeholders as possible, and to develop objectives that can be accepted and adopted by all those with an interest in the use and survival of the area concerned. These guidelines, based on good practice in many parts of the world, provide a framework for protected area planners to consult and adapt to their own needs and circumstances.

These guidelines are intended to update and extend the guidelines published in 1980, and to reflect developments and new issues that have arisen since then. These developments include important advances in international environmental law, as well as greater scientific understanding of the role of protected areas in nature conservation, which includes conserving biodiversity, maintaining ecosystem functions and contributing to sustainable development. Like the first version in 1980, these new guidelines for protected areas legislation are aimed first and foremost at drafters of legislation working closely with protected area authorities, as well as other players in the legislative process. They will also be an important resource for agencies responsible for monitoring and implementing policies and programs that affect protected areas legislation, or vice versa.

The development of a Management Plan can be more or less complex, depending on the objectives of the protected area, the risks and threats affecting these objectives, the number of interests involved, the level of stakeholder involvement and the problems posed outside the protected area. Whether the plan is simple or complex, reasonable planning principles should be applied to guide the planning process and ensure that the Management Plan as a whole is a clear and useful document. These Guidelines, based on good practice used in many places around the world, provide a framework for protected area planners to consult and adapt to their own needs and circumstances. These management planning guidelines have, in general, been written in the context of management by a central, provincial or local government body.

3.2.. Management action

The main role of a (protected) site manager is to manage the natural or cultural heritage of a protected or specially managed area. He or she oversees the protection and management of this area, as well as the promotion, awareness-raising and mediation of this heritage. He or she coordinates observations and interventions with a view to preserving the site's heritage quality   and   welcoming the public.

He or she is the guarantor of the proper management of protected areas. To this end, he or she is increasingly developing a territorial approach, with a management strategy that extends beyond the site. He or she can thus implement actions outside the strict perimeter of the protected area.

When the person in charge of the management and conservation of the protected site is also the person with administrative and strategic responsibility for it, the job is then to be completed with the "site responsibility" function sheet.

3.2.1. Main tasks and activities:

- **Coordinate the site's conservation missions and activities:**

Organize work and monitor progress, ensure compliance with company procedures, safety rules and regulations,

Manage a multi-disciplinary team and distribute work within it,

Organize and coordinate environmental monitoring and policing activities, in conjunction with the relevant departments (AFB, ONCFS, etc.).

- **Implement the management plan :**

Support the implementation of the management plan through the programming of an action plan with the technical team,
Plan and monitor work,

Set up the scientific framework for implementing the management plan (scientific monitoring, data collection and analysis),

Carry out and/or participate in field missions (inventories, mapping of protected species, scientific monitoring, various work sites, etc.), oversee the evaluation of the management plan, carry out the assessment, and its renewal.

3.2.2. Contribute to the strategic positioning of the site and structure :

Contribute to forward-looking thinking, make strategic proposals for site management,

Develop joint projects with the various institutions and communities involved in the site,

Contribute its expertise to institutional players and partners, Involve its site and management activities in networks (networks of protected areas, scientific and cultural networks, networks for exchanging practices),
Represent the organization at institutional meetings.

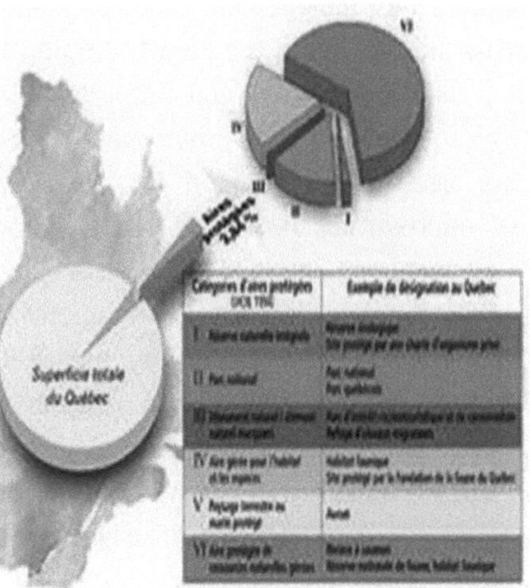

Within a given PA, there are likely to **be** areas where the planning team will decide to target different activities or emphasize different aspects offered by the protected area. These areas, known as microzones, should be delineated on the PA map, and their characteristics defined in this section of the plan. For example, if a tourist center is planned for a given area, there will be a greater concentration of people in that area. This type of micro-zone could be called a nature discovery micro-zone, and the planning team may decide to target future development within it. Other micro-zones can be planned as buffer zones between areas of high human activity and areas that need to be protected from human disturbance. In the accompanying text, provide a title for each micro-zone and explain how it will be managed and why it warrants being managed differently, define the objectives set for the zone and describe the guidelines for the micro-zone. Micro-zones are not used to describe places with different ecological characteristics, but rather places with different management actions. For example, if an entirely forested PA nevertheless contains a patch of savannah, this area will need to be described as a micro-zone unless the management guidelines for this zone differ to some extent from those for the rest of the PA.

For each micro-zone, the accompanying text should name the micro-zone and explain the management objectives and guidelines specific to that micro-zone. To avoid the plan becoming too complex, and to ensure that it is implemented and enforced more easily, keep the number of micro-zones to a minimum, and remember that the entire PA does not need to be divided into micro-zones. Reserve the allocation of micro-zones for places that need special protection or more rigorous management than the PA as a whole. To determine whether it is necessary to include a new micro-zone in the plan, the planning team will need to consider how management in this specific location will differ from management throughout the rest of the PA or in any of the other micro-zones. Remember that the guidelines applying to the whole PA will always apply within the micro-zones, unless the micro-zone guidelines expressly provide for the exemption of certain restrictions. The guidelines in place for the PA do not therefore need to be repeated for each micro-zone.

Recreational fishing: for tourists Heritage or cultural sites Developed or high-impact areas: for administrative buildings, B&Bs, etc. Extractive use zones: where the extraction of limited quantities of timber or non-timber products may be permitted in accordance with certain guidelines Village zones: whether villages exist within the PA itself and whether they will be allowed to remain.

Different PAs will need different micro-zones to achieve all their desired conditions and objectives. Some micro-zones to consider for a PA plan are listed below (again, keep their number to a minimum; if there is no appreciable difference in the guidelines of two different micro-zones, it may be preferable to combine them into a single micro-zone): Complete protection zones: places entirely off-limits to access, perhaps with the exception of park administration or limited research Hunting grounds: where certain communities have the right to hunt within certain guidelines that are specified in the plan Subsistence fishing: reserved to

3.4. Implementation

A management plan is a document that sets out the management approach and objectives, and includes a framework for decision-making, to be applied in the protected area over a given period. It is now critical for protected area planning to consult as many stakeholders as possible, and to develop objectives that can be accepted and adopted by all those with an interest in the use and survival of the area concerned. These guidelines, based on good practice used in many parts of the world, provide a framework that protected area planners can consult and adapt to their own needs and circumstances.

The development of a Management Plan can be more or less complex, depending on the objectives of the protected area, the risks and threats affecting these objectives, the number of interests involved, the level of stakeholder involvement and the problems posed outside the protected area. Whether the plan is simple or complex, reasonable planning principles should be applied to guide the planning process and ensure that the Management Plan as a whole is a clear and useful document. These Guidelines, based on good practice used in many places around the world, provide a framework for protected area planners to consult and adapt to their own needs and circumstances. These management planning guidelines have, in general, been written in the context of management by a central, provincial or local government body. However, we can adapt them to suit specific needs and circumstances. These management planning guidelines have, in general, been written in the context of management by a central, provincial or local government body.

3.5. References

A commitment to an ambitious, interconnected and effective site-based conservation network, representing all areas important for biodiversity and ecosystem services, is of paramount importance. Such a network must take into account the custodial roles and responsibilities assumed by indigenous peoples and local communities, and recognize that their different uses of biodiversity can be compatible with effective conservation.

- Marseille Manifesto

We work to establish practices, standards and policies that maximize the effectiveness of protected and conservation areas, promote the sustainable use of landscapes and advance conservation justice and equity. As the official nature advisor to UNESCO's World Heritage Committee, we evaluate natural World Heritage candidate sites and monitor those already recognized as the world's most important protected areas.

3.6. PA planning process

A protected area management plan describes the actions needed to ensure that the protected area serves the purpose for which it was created. Planning is the process by which stakeholders (community members, scientists, government representatives, private companies, etc.), come together to debate and consider how to manage the land for the benefit of present and future generations and ensure the ecological sustainability of the land and resources. The plans establish guidelines and objectives for the PA over a specific period of time, regardless of changes in personnel.

PA planning can be problematic if the issues inside or outside the area are complex. Planning involves risk assessment and forecasts of anticipated and uncertain future events and conditions. As a result, even the best plan will need to be modified to adapt to improvements in data and information; changes in social, economic or other conditions; the progression of threats; the role of the PA within the wider landscape, or the results of monitoring efforts. As a result, plans are adaptive in nature, and modifications or complete revisions can be made at any time.

Their nature and complete modifications or revisions will result from PA monitoring and other factors examined in the plan.

At the planning stage, it is essential to recognize that not all data concerning the PA and its resources will be available at the desired level of detail. This is true for all PAs worldwide, regardless of the financial and human resources available to the managing authority. Nevertheless, PA planning must take place with the understanding that the plan may require the collection of specific data, and may need to be revised in the light of newly acquired data in order to make better-informed decisions. Consequently, it is important not to delay the development of the plan for lack of data.

AP plans around the world vary considerably in content, detail and complexity. As the planning process unfolds, it's worth remembering that simpler plans are often more effective plans. The chances of the plan being widely read and understood by local stakeholders, as well as the chances of their engagement in the process, will be greater if the plan is concise, focuses on what is important to protect the PA and the resources it contains, and is expressed in clear language, both scientific and legal. Such an approach will also facilitate implementation of the plan.

3.6.1. Protected area planning process

 The following sections outline the elements of the AP planning process, including the elements of AP plan development, how these elements can be fleshed out, and other considerations of the planning process and plan development. In addition, many of these sections include "tasks", i.e. actions expected of CARPE's implementing partners and used by USAID/CARPE as monitoring tools to measure progress.

 Protected area planning in the context of CARPE The PA planning process should demonstrate how the community, CARPE implementing partners and other relevant stakeholders will have: 1) assessed and analyzed activities, resources, uses and trends in the PA in question; 2) developed and formulated objectives and desired conditions for the PA; 3) consulted, collaborated with and considered stakeholders in the development of the plan, and 4) targeted management activities to achieve the desired conditions and priority objectives with the participation of relevant stakeholders.

Although it may vary to a greater or lesser extent at national level, in Central Africa the typical life cycle of a plan will usually be 5 to 10 years, with annual monitoring and evaluation. This monitoring and evaluation will determine whether conditions or needs have changed sufficiently to warrant updating the plan, or whether the assumptions made during the planning process are correct.

These plans are required as part of CARPE program activities and are intended to foster collaboration between HAs, focus efforts on priorities and stimulate planning processes throughout the region.

3.6.1.1. Stages in the planning process

The following steps form the basis of the PA planning process: 1. Identify the planning team and define roles 2. Analyze existing legislative process for PA management plan approval 3. Collect data a. Define the PA's resource characteristics and conditions (this step involves a synthesis of current knowledge about the PA and its surroundings; the text on these PA characteristics should be limited and precise; the plan is not a research document) b. Demarcate the PA (this step should take place during the macro-zoning process as part of landscape planning; however, it may be necessary to define the PA's boundaries more precisely) c. Identify other stakeholders (groups the planners may not initially know about, and groups with an interest in the PA who do not reside in the immediate area) d. Assess the PA's legal status (proposed, recognized, neither) e. Identify trends in resource conditions and use by local populations.

3.6.1.2 Planning identification and role definition

f. Identify significant information gaps 4. Specify methods of public participation 5. Define desired conditions and objectives, describing desired conditions for the PA 6. Define resource use guidelines for the entire PA 7. Identify micro-zones and regions requiring special management, and define objectives and guidelines for each 8. Obtain official approval or endorsement of the plan 9. Implement the plan 10. Monitoring and evaluation

11. Revise and update the plan as information improves, conditions change and monitoring yields results.

This description will include a review of the roles and responsibilities of the various parties involved in implementing the plan; the public participation strategy; the plan's monitoring and evaluation approach; and a multi-year implementation schedule that will present the planned management action program to facilitate more detailed planning of the work.

This section should specify the different roles and responsibilities of public bodies and other organizations in administering the plan. Various institutions will be responsible for actions associated with plan implementation, such as project design and approval, project execution, budgeting and monitoring. Execution of the plan's management actions will need to comply with the country's legal framework, which includes applicable laws and regulations and follows specific protocols administered through government institutions. In this section, describe who will be responsible for each action to implement the plan.

Due to the limited management capacity and resources of the ministries that are the authorities by law, the actual management of the PA and its resources will often fall to a range of actors present in the landscape. Government ministries and departments, conservation organizations and other NGOs, the private sector and local communities will bring a range of capacities and resources to help implement the plan. Because of this uncertain nature, it will therefore be important to create (or strengthen) the relevant management and advisory teams, and to assign responsibility for direct implementation aspects of the plan to individuals or organizations with the necessary skills and resources to take them on.

3.6.1.3. Analyze the existing legislative process for approval of the PA management plan

Making laws is one of Parliament's greatest responsibilities. So it's hardly surprising that the legislative process occupies a large part of parliamentarians' time. The legislative stages described below are the final links in a much longer process that begins with the proposal, formulation and drafting of a bill.

In the case of government bills, drafting is entrusted to the Ministry of Justice, which follows Cabinet directives.

Non-Cabinet members who are not Parliamentary Secretaries may introduce bills for consideration under Private Members' Business. These bills are usually drafted on behalf of an MP by a legislative counsel employed by the House.

3.6.2. Types of bills

There are two main categories of bills: public bills and private bills. Public bills deal with matters of national interest, while private bills grant powers, special advantages or exemptions to one or more persons, including corporations. Most bills considered by the House of Commons are public bills.

Government bills" are public bills introduced by ministers, and "private members' bills" are bills introduced by private members.

3.6.2.1. Collecting data

A four-phase process

(1) Preparation

This first stage involved setting up a multi-disciplinary planning unit made up of DCRHM management staff, representatives of local communities and civil society, and experts from the Technical and Scientific Department. A total of 35 people were involved in this unit.

(2) Data collection

Data collection was carried out in a participatory and inclusive manner, in compliance with the **World Commission on Protected Areas**' guidelines, rules and principles. A first series of data was collected during workshops and discussion groups involving the various stakeholders in the territories.

Next, a wildlife survey was carried out within the protected area to identify its wildlife and cultural values.

Once the data had been collected, the team from the General Management's Technical and Scientific Department and the site planning unit, in collaboration with Kinshasa, proceeded to analyze the data and draft the plan. The aim of involving site staff in this way was to make them co-responsible, and to bring collective intelligence to the fore for the benefit of the department in particular, and the Ministry of the Environment in general.

As the reserve has no up-to-date data, this exercise proved fascinating, as everyone was keen to learn more about the state of wildlife in the area. These exchanges prompted the experts to reflect and make strategic choices concerning the reserve's development interventions and the formulation of management plan objectives.

In addition to conservation areas, the mapping exercise identified critical or highly degraded areas, traditional use areas or areas with little degradation, artisanal timber harvesting areas or areas formerly exploited by logging companies, and tourist areas.

Other sources of information from study products or work being carried out simultaneously in the landscape by (project partner) were then used to deepen these analyses. These included reports on the establishment of the governance working group, the land use dialogue survey report and the landscape restoration opportunity assessment.

(3) Local validation

Members of the governance working group represented 20% of participants. Local communities were represented by natural resource user associations. Around 49% were farmers, 14% hunters, 9% timber harvesters and 8% fishermen. Other stakeholder groups present included teachers, religious leaders, land and customary chiefs, young people and health personnel (34%).

Using an evaluation sheet on the degree of use of good and bad natural resource management practices in the landscape, this workshop was also an opportunity to generate other types of data and assess the project's influence on the attitudes and behaviour of local people. This assessment was carried out in a participatory manner within each administrative sector concerned by the protected area. Overall, it showed that the use of natural resource management practices had changed in very few administrative sectors. Although people are aware of Project PLUS thanks to awareness-raising activities, there has not yet been much influence on the behaviour of natural resource users to abandon bad or adopt good practices.

(4) National validation

The development and management plan was then presented to other stakeholders at a workshop held in Kinshasa in December 2019. This validation workshop was attended by national deputies, territorial administrators and representatives, representatives of non-governmental organizations involved in the landscape, and the site manager. The MPs and administrators present praised the relevance of the management plan and commented on the naming of certain villages and sector boundaries before proceeding with its validation.

The presence of deputies from the National Assembly at the workshop gave it added significance for having involved the DRC's legislative power players. The participants' comments were incorporated by a small team of experts. The full document is available and was presented to the

Direction Générale d Thus, the DCRHM is the sixth protected area in the DRC to have a development and management plan validated at local and national level by all stakeholders.

3.6.2.2. A plan for the future

The community-led governance working group is specifically recognized in the management plan as being responsible for governance of the protected area at the local level, thereby increasing local power and enabling people to enforce their rights across the landscape.

The success of the process that led to the development of this management plan may influence others in the DRC's network of protected areas. Several national parks and reserves are in the process of developing or revising their management plans. The lessons learned from the process will serve as a valuable model that can be duplicated in other protected areas.

3.6.3. Define the role of the planning team

The role of the strategy development manager can be divided into the following three functions:

1. To guide Board members and management in their decision-making on new directions or areas of development;

2. Take charge of or support in-house or partner consultations;

3. Carry out a detailed analysis to identify opportunities based on internal data or data from the organization's own environment (competition, customer needs, social changes and economic environment);

Once the strategy has been approved, direct or supervise the execution of specific projects.

Whether it's a committee or an individual designated for this role, a clear understanding of these functions helps clarify the expectations of the Board of Directors who will assume final approval of the strategic plan.

a) Main responsibilities and tasks

This role involves responsibilities and tasks, the main ones of which are described below. The scope of each may vary according to the size and current situation of the organization, and the associated risks and issues (e.g. financial, continuity, technology, human resources, social responsibility and ethics).

1. Work with senior management to determine strategic objectives in a collaborative and inclusive process, and develop action plans to align medium-term objectives with long-term strategy;

2. Assess existing systems and the need for improvements, as well as the need to modify structures, processes, practices and assessment methods;

3. Document data on the requirements and contributions of the various partners, to support the updating of agreements and strategies with each of them;

4. Establish the need to implement business intelligence tools or databases enabling decision-makers to have access to relevant information and thus enable optimal performance monitoring and reporting;

5. Specify the various projects to be carried out in order to create the required changes;

6. Design a communications plan that conveys the new strategy, upcoming initiatives and anticipated results with impact;

7. Prepare a presentation (and/or report) for submission to the Board on the proposed strategy, the rationale for the choices made, and the action plan with timetable.

b) Desired skills and experience

These responsibilities and tasks are in line with the general approach used to carry out strategic planning. To ensure success, strategy development requires a combination of skills and experience on the part of the project leader or assigned committee.

On the **education** side, we usually ask:

• a university degree, or equivalent training or professional experience and/or a professional designation.

For **the experiment**, we are looking for :

- successful management/implementation of comparable projects in the same environment, or development of growth plans, merger projects or management consulting;

1. experience of the environment, with Document data on the requirements and contributions of the various partners, to support the updating of agreements and strategies with each of them;

2. Establish the need to set up business intelligence tools or databases enabling decision-makers to access relevant information and thus enable optimal performance monitoring and reporting;

3. Specify the various projects to be carried out in order to create the required changes;

4. Design a communications plan that conveys the new strategy, upcoming initiatives and anticipated results with impact;

5. Prepare a presentation (and/or report) to be submitted to the Board of Directors on the proposed strategy, justification of choices, and action plan with timetable.

Document data on the requirements and contributions of the various partners, to support the updating of agreements and strategies with each of them;

1. Establish the need to implement business intelligence tools or databases enabling decision-makers to have access to relevant information and thus enable optimal performance monitoring and reporting;

2. Specify the various projects to be carried out in order to create the required changes;

3. Design a communications plan that conveys the new strategy, upcoming initiatives and anticipated results with impact;

4. Prepare a presentation (and/or report) for submission to the Board

on the proposed strategy, the rationale for the choices made, and the action plan with timetable.

c) Desired skills and experience

These responsibilities and tasks are in line with the general approach used to carry out strategic planning. To ensure success, strategy development requires a combination of skills and experience on the part of the project leader or assigned committee.

On the **education** side, we usually ask:

- a university degree, or equivalent training or professional experience and/or a professional designation.

For **the experiment**, we are looking for :

- successful management/implementation of comparable projects in the same environment, or development of growth plans, merger projects or management consulting;
- experience in the field, **skills/abilities,** consider :
- Accomplished, proactive leadership, agility, confidence;
- Strong, articulate communicator who excels at acting with vision and collaboration;
- Ability to interact with partners, and initiate discussion on strategy;
- Good judgment, ability to synthesize and find solutions to problems.

There are many tools your management consultants can use to recruit the right candidates. Taking the time to describe the profile of the ideal manager will help you frame the actions and the strategic planning process.

3.7. Stakeholder participation

By participatory management or co-management of protected areas we mean a form of partnership enabling the various stakeholders to share the function, rights and responsibilities relating to the management of a territory or range of resources enjoying protected status.

The term "participative management" refers to the same thing as "co-management" or "joint management". Multi-partner management or joint management agreement, describes a form of partnership.
In which all interested parties agree to share management functions, rights and responsibilities over a portion of territory or resources.

Participatory management implies recognizing the legitimacy of communities in the management of the natural environment. Stakeholders are aware of their role in the management of the protected area, and have the knowledge and skills to contribute to management.

3.8. Guiding principles

> ➢ Those affected by a decision can influence its content
> ➢ In principle, participation increases the quality of decisions
> ➢ Sharing decision-making power increases players' loyalty to this authority
> ➢ The time invested in decision-making is offset by the shorter time required for implementation
> ➢ Priority is given to the group as the decision-making unit
> ➢ Managers remain fully responsible
> ➢ Open-mindedness
> ➢ Respect for people
> ➢ transparency
> ➢ mutual trust
> ➢ faith in people's abilities and expertise.

3.9. Stages in the development of a management plan

The steps involved in developing a development plan are :

Form the planning team: this should include people with experience in planning, ecology, sociology, economics and other scientific fields. It should also include the protected area authorities and the people who manage it, as well as those who will be affected by the plan. The team should consult with scientists, educators, tourism experts, concession owners and people living in and around the protected area.

Establish the objectives of the protected area: list and analyze the original reasons for creating the protected area, and if necessary update the objectives to reflect current conditions. Wherever possible, prioritize these objectives.

Gather existing information: this should include a list of current legislation, as well as biophysical, cultural and socio-economic data.

Start compiling information in an organized way so that it's easy to find when new decisions need to be made.

Obtaining new information: once you have compiled the existing information, update it as necessary with new data, using, for example, the techniques presented in this manual. Chapter 2 provides information to search for. Make sure that the database design allows for the easy inclusion of new information.

Estimate management constraints: limitations of an environmental, economic, political, administrative or legal nature must be recognized and analyzed. It's important to be realistic.

Study regional interrelationships: the protected area must be an integrated element of land use, and the management plan must be designed accordingly. Consider the potential effects of outside development on the resources of the protected area, as well as the impact of the protected area on the region. Study the boundaries of the protected area: consider boundary modifications based on ecological, cultural, socio-economic or administrative factors.

Determine appropriate management zones: determine the management intensities in different zones, differentiating for example between a strictly protected "core zone" and a buffer zone with multiple uses, etc.

Develop detailed action plans: once overall agreement has been reached on new objectives for the protected area, outline precisely the action plans needed to achieve the objectives. Action plans should describe in detail who will undertake the work, and how it will be done. There should be plans for each of the following categories: resource management and protection; human use; research and monitoring; administration.

Integrated action plans to create integrated development options: study the infrastructure and financing needed to accomplish the various programs.

Highlight financial requirements: determine the costs of different planning proposals and identify potential sources of funding.

Prepare and distribute a draft plan for comment.

Analyze and evaluate the plan based on feedback from stakeholders.

 Draw up action plans and schedules.

Prepare, publish and make public the finished plan.

Monitor and revise the management plan and activities regularly, based on practical experience.

3.10. Clarifying concepts

3.10.1. Stakeholder

A stakeholder is an individual or collective actor (group or organization) who may be affected by or interested in the management of a protected area, even if they are not directly affected by the impact of protected area activities.

Generally, the planning team must use methods to involve stakeholders in the planning and management of the protected area to achieve good results, and also ensure that the views of stakeholders are incorporated into the protected area plan at every possible level.

By taking stakeholders' views into account, creating a sense of empowerment among local community members, and broadening the audience of stakeholders by involving them in planning discussions and decisions, we improve the chances of success of the plan and its implementation.

So to improve the likelihood of long-term success, it is essential that local communities derive tangible benefits directly from the existence of the protected area.

3.10.2. Prioritization of planning actions

Task prioritization can be defined as planning the actions in your diary in order of urgency and importance. Prioritizing tasks makes you more productive.

3.10.3 Sustainable financing

Sustainable financing is the set of financial activities aimed at improving the interests of the community, entity or protected area over the long term.

3.10.4 Objective

The aim of this work is to enrich and complete our training in protected area management.

3.11. Work methodology

To carry out this work, we read a number of available documents, including books and magazines, and consulted a number of websites.

3.12. Development

3.12.1. Protected Area (PA) planning process

A management plan guides and controls the management of protected area resources, the uses of the area and the development of infrastructure necessary for management and use. It determines the development of activities and management actions to be carried out in an area, and details the mechanisms for assessing the success or failure of these activities. At the heart of the plan is a statement of measurable goals and objectives to guide the work of managers. These goals and objectives form the framework for determining what actions will be taken, when they will be taken, and the budget and personnel required. The plan must describe the need for specialized training, which ensures that the team is competent and knows when to make changes. A development plan is written for a specific period, usually five years, at the end of which it is evaluated and modified. Annual work plans are developed during the execution phase, guided by the long-term plan.

The management plan should remain subject to modification as new information becomes available. It should be particularly sensitive to the evaluation of actions undertaken in previous annual plans.

3.12.2 Stakeholder identification

Stakeholder participation will first require the planning team to identify the stakeholders and the methods of exchanging information between them and the team. The following points should be taken into account when involving stakeholders:

Names of key participants, specifying the stakeholder group(s) they represent.

Identify groups who are key to land-use decisions, who have an impact on the PA, or who benefit from the resources it supports. During the identification process, seek to reach out to those who work in a field unrelated to natural resources and who can help provide useful information, or who may know of other individuals or stakeholders involved (e.g. health workers or teachers who may know of individuals or organizations that could contribute significantly to the landscape planning process). Include representatives of central and provincial government, as well as traditional authorities, among stakeholders.

Consider including ethnic and religious groups, timber harvesting companies, tourism companies, and other stakeholders.

mining companies, NGOs (local, regional and national), public bodies, civil society, hunters, fishermen, loggers, farmers, water users, researchers and other groups with an interest in the PA.

Take into account activities carried out outside the PA that may have an impact on its resources, identifying the groups or individuals overseeing these activities.

Are there any proposed developments or infrastructure in the PA, such as road repairs? Who oversees these activities and makes decisions on road location?

Given the PA's priorities and trends, decide which stakeholders are essential to the PA's priority review and planning decisions. Are any stakeholder groups threatening the HA's essential resources?

Is there a risk of conflict between the interests of some of the stakeholders?

are there any influential authorities operating or living in or around the PA?

What stakeholder interests may oppose micro-zoning decisions?

These questions can help to identify stakeholders and determine the order in which their intervention is prioritized.

Who are the stakeholders in a protected area?

A PA's stakeholders will vary according to the history, resources, socio-economic conditions and other aspects of the PA and its surroundings. Stakeholders in a given PA may include :

a. Villagers living in or near the PA boundaries

b. Communities further away from the PA that depend to some extent on its resources, or that pass through it. c. Traditional authorities.

d. Government representatives at national, regional and local level

e. Excluded groups who can't always express themselves in the above groups

f. Individuals claiming ancestral rights to the land

g. Extractive industries, whether operating within the PA itself or outside its boundaries h. Local NGOs

i. The international community and NGOs

k. The tourism sector, etc.

3.12.3. Stakeholder participation strategies

The PA planning process will involve a range of stakeholders, with different levels of participation. It may be necessary to use different strategies to involve stakeholders. Identify methods of exchanging information between stakeholders and the planning team. The following points should be included or taken into account when involving the various stakeholders:

Determine how planning staff will communicate with stakeholders (i.e. individual and/or group meetings in the landscape and/or in a central location) and specify which stakeholder groups, if any, will be treated differently and why.

Consider whether all stakeholders will be able to commit sufficient time to participate in the HA planning process. If they are unable to attend organized group sessions, and if their participation is essential to the success of the planning process, consider keeping them informed through personal communications.

Explain how information will be exchanged and how concepts will be communicated to the various stakeholders. This is particularly important for local residents, as many have little or no access to maps, data and reports, and some may have a low level of education.

Define the general purpose of each communication with stakeholders; e.g., information exchange, data gathering, decision-making, etc.

Consider how stakeholder representatives will coordinate between the PA planning team and their respective groups to ensure that information and views are properly conveyed and received. Define specific topics of conversation for each stakeholder group and for conveying concepts to the group as a whole. Include well-defined terminology to reduce confusion during the planning process.

Indicate which languages will be used for written documents and oral communication, and how the planning team will meet translation needs.

Ensure that all participants understand the planning process and their role in it.

It is desirable to achieve a high level of stakeholder participation during the planning process.

***Stakeholder engagement**

The commitment of the players involved is particularly important in

The environment, where the interconnection of aspects relating to the sea means that actions in one area have an impact on another. Partnership with local communities is also justified in terms of the legitimacy of several common management interests, such as the exploitation of traditional fisheries.

Participation builds trust and confidence between the parties involved, and helps to build consensus. It ensures that local perspectives are understood, local concerns are addressed and local know-how is utilized. Stakeholder involvement mobilizes and strengthens local capacities, prevents conflict and generates a collaborative social climate, making conservation efforts more effective, useful and sustainable. In P.A.s, participation helps build a shared vision with local stakeholders regarding conservation and sustainable development. It encourages local collaboration in protecting the area and confers a sense of pride and ownership over the site. Participation can mean the difference between strictly protecting a P.A. behind its fences or integrating it into local culture, customs and regulations, encouraging the creation of a sustainable society all around.

Seen from the angle of sustainable development, participation brings the social dimension to economic objectives and ecological concerns relating to the use of natural resources. It is essential from a sustainability perspective. Integrating it into the management of protected areas is a test case for sustainable development, particularly through the natural resource management transfer programs set up by conservation stakeholders.

Benefits of participation.

• It enables the development of a shared vision and commitment to a common goal.

Common

• It enables the identification of shared priorities and realistic actions

• It builds trust between different groups and prevents conflict.

• It gives legitimacy to conservation processes

• It enables local viewpoints to be understood and integrated - It makes use of local knowledge and provides new sources of information

• It uses and energizes existing local capacities

• It creates positive synergies

• It strengthens local capacities and creates social capital

• It improves the efficiency of agreed activities

• It reinforces the stability, continuity and sustainability of the process

Participation does not replace the decision-making process, but helps to set it up and contributes to its success. Planning and management must assess which issues should be addressed only at national and central level, and which at local level. To this end, stakeholder involvement does not mean that the relevant authorities delegate or lose their decision-making powers or responsibilities.

3.12.4. Prioritization of planning actions

While it's ideal to invest a great deal of effort in each stage of the planning process, as well as in implementation and monitoring activities, the reality of limited financial and human resources, as well as many other business difficulties in the region, will prevent planning teams and authorities from achieving these perfect levels of planning action.

Therefore, it is important for the planning team to follow a prioritization process throughout planning, execution and monitoring. The essential steps for effective resource prioritization are data gathering, plan execution and monitoring. Honest assessments of available funds and the cost of specific activities will help determine what the planning team can realistically afford to accomplish. The planning team and/or the PA administration must determine the major threats facing the PA, the opportunities the PA offers, and assess the partner organizations that can or might complement the actions taken by the PA administration.

3.12.5. Sustainable **financing**

Financial needs must be emphasized here: determine the costs of the various planning proposals and identify potential sources of funding.

The defined plan should lead to the development of a business or financial plan that specifies the sources of potential revenue, specifying partnerships, budget, cost sharing and the release of funds to help implement the plan, funds to be used in particular for operations and fundraising campaigns.   *Overall objective

Ensure the long-term financial viability of national PA systems in a country. Establish capacities, institutional frameworks and model mechanisms for the long-term financial viability of PA systems.

*Source of financing

From entrance fees to tourist concessions: a circuit that excludes local populations;

Entrance fees are a common source of funding for protected areas.

Opportunities in the fight against climate change and carbon funds. External financing: foundations/debt-nature/trust funds: Among external financing, we will first consider foundations, debt-nature swaps and *trust funds*. The current worldwide trend is to combine these three instruments into a single sustainable financing tool.

In partnership with :

Generally speaking, it must have partners who have worked with it who have the financial means to meet the objectives of the management plan to conserve biodiversity in an ecosystem (PA).

However, the final plan will need to lead to the development of a business or financial plan that specifies potential sources of revenue, outlining partnerships, budget, cost-sharing and the release of funds to help implement the plan, funds to be used in particular for operations and fundraising campaigns.

Conclusion

It is said that "the end of a thing is better than its beginning", and the fire will burn if everyone brings a piece of wood. By undertaking this work which marks the end of our research on the subject entitled: De la question approfondie d'aménagement; Cas de la République Démocratique du Congo by the author SHONGO YONGA Jean, associated with his aforementioned comrades in arms under the supervision of the ordinary professor, **LOHAKA Djonga.**

We have set ourselves the goal of describing and analyzing the planning of a protected area in **DR Congo**, as in the rest of the world.

Our concern in dealing with this subject was to :

- provide background and define protected areas,

- know how to manage a protected area,

- plan a protected area.

After providing explanations to the points planned for this work, we suggest to all our readers and those called upon to manage protected areas in DR Congo to follow the laws:

- Law no. 11/0009 of July 09, 2011 on the fundamental principles of environmental conservation.

- Loi n° 14/003 du 11 février 2014 relative à la conservation de la nature, De la République Démocratique du Congo.

BIBLIOGRAPHY

1. Bertrand Chardonnet 2014: Consultation for the improvement of the ecological monitoring system in the protected areas of Ivory Coast. 177p. World Heritage Committee 2013:

2. GTZ, (1983) - Mosso-Buyogoma Master Plan. Regional Planning for the Provinces of Cankuzo, Rutana and Ruyigi in Burundi. Tome I and tome II.

3. ICCN, Manuel des Droits et Obligations des Parties Prenantes dans les Aires Protégées, May 2011, DRC

4. IUCN. 1994. Guidelines for Protected Area Management Categories. CNPPA with the assistance of WCMC. IUCN, Gland, Switzerland and Cambridge, UK. x + 261pp.

5. IUCN-WCMC Guidelines for Protected Area Management Categories (http://www.unepwcmc

6. Law no. 11/0009 of July 09, 2011 on the fundamental principles of environmental conservation.

7. Law no. 14/003 of February 11, 2014 on nature conservation

8. Malaisse, F. (1979) - The miombo ecosystem. In: Ecosystèmes forestiers tropicaux. un rapport sur l'état des connaissances UNESCO, PNUE et FAO: 632-657.

9. Ministry of Development Planning and Reconstruction Nationale (2006)- Monographie de la Province de Rutana. Local Planning Support Project (PPL)/UNDP,

10. Ministry of Forests and Fauna (MINFOF), Cameroon. 2008. Directives pour l'élaboration et la mise en œuvre des plans d'aménagement des aires protégées du Cameroun. Yaounde, Cameroon.

11. Rapport de mission sur l'état de la conservation du Parc de la Comoé. 13p. IUCN 2013: Monitoring law enforcement in protected areas: Necessary for conservation, but insufficient for good governance. NAPA N° 64, protected areas news. 10p

12. Sind Akira, A., (2007) - Environmental Impact Study. Project Report

Table of contents

More
Books!

info@omniscriptum.com
www.omniscriptum.com
OMNIScriptum

Printed by Books on Demand GmbH, Norderstedt / Germany